Jack and the Girl Who Believed in Him

Book 2 of the 5 Daughters Ranch Series

Donna Minton Williams

Minton Legacy Publishing

Jack and the Girl Who Believed in Him
Book 2 of the 5 Daughters Ranch Series

This is a true story. Some details have been gently shaped for young readers.

Published by Minton Legacy Publishing
ISBN (print): 979-8-90523-022-6
ISBN (ebook): 979-8-90523-023-3

First Edition, 2026
Printed in the United States of America

For Bryer.

The fearless little cowgirl who climbed the fence to look a great black horse right in the eye, and loved him just the way he was.

You saw Jack not for his one eye or his old age, but for the beauty he carried inside. May you always see the world that way, and may you never stop believing.

The day you stop dreaming is the day you stop living.

Contents

The Horse Nobody Wanted 1

The Head That Came Down 10

Learning to Trust Again 17

The Language Between Them 24

Buttons and Bruises 30

The Summer at A&M 36

Coming Home Different 43

The Horse Who Kept Her Secrets 48

Bryer's Turn (and Everybody Else's) 56

When Running Hurts 61

The Last Bath 68

Heaven's Pasture 75

What Jack Left Behind 80

Chapter 1

The Horse Nobody Wanted

Jay kept both hands steady on the worn leather wheel as the old truck rumbled down the dusty country road. The Texas sun hung high and fierce, throwing a golden glare across the cracked asphalt and baking the earth a deep, rusty red. Dry mesquite and sagebrush drifted through the open windows, the same warm, dusty smell that always clung to the ranch roads after a long, hot day.

Beside him, Donna sat quietly. Her hands rested in her lap, but her heart would not slow down. She found comfort in Jay's quiet strength, in his broad shoulders and his set jaw. He was a man used to weathering storms, the kind in the sky and the kind in life.

Back home at the 5 Daughters Ranch, their littlest granddaughter, Bryer, was barely a year old. She was far too small to know that anything important was happening that day. She would not remember this drive, or the sad place at the end of it, or the long, hard work that was about to begin. Someday, when she was older, the grown-ups would have to tell her the story. For now, that story belonged only to Donna and Jay and the wide Texas sky.

Ahead, the horizon stretched empty, a pale blue sky meeting endless stretches of scrubby brush and scattered oaks. The trees stood like lonely sentries in the heat. Cicadas buzzed without stopping, and somewhere in the tall, sun-bleached grass, a meadowlark called out.

Donna barely heard any of it. Her mind was tangled up in memories, sharp and aching, of a certain black horse with one eye.

Jack.

She could hardly believe it was the same horse. Not long ago, Jack had been the pride of a fancy, high-dollar ranch. His sleek black coat had gleamed in the sunlight like polished glass. She could still picture the way his muscles rippled as he carried a teenage girl across the arena, smooth and powerful, graceful as a dancer.

Jack was a champion. People said he had "automatic buttons." The smallest signal, a gentle squeeze of a leg or a soft shift of weight in the saddle, and he would spin in a circle or stop on a dime. Donna had ridden him many times back then, always amazed by how smart he was, how quietly strong. He seemed to understand every command before she even gave it, as if he could read her mind.

He was more than a horse. He was a partner. A friend.

Now the truck tires crunched over gravel as Jay slowed down near a small, run-down property. This was the place where Jack and a paint mare named Cheyenne were being kept.

Donna's heart hammered, each beat louder than the last.

She had heard the rumors. Whispers from the neighbors. Hushed talk at the feed store. People said the horses out here were being starved, left to waste away by someone who was supposed to care for them. But nothing could have prepared her for what she was about to see.

Jay eased the truck to a stop beside a weathered fence, the wood gray and cracked from years under the brutal Texas sun. The paddock beyond was parched and dusty, the ground hard and split like an old, dry riverbed. Not one blade of green grass remained. The dirt was a dull brown, scattered with brittle leaves and broken twigs. The air smelled sour and warm, a faint hint of manure mixed with sun-baked earth.

Two horses stood near the far gate. Their thin bodies stood out sharply against the bare land. Their ribs jutted under dull, matted coats that had once shined with health.

One was Cheyenne. Her bright white and brown patches had faded and sunk in.

But Donna's eyes went straight to Jack.

The glossy black coat she remembered was now a patchy shadow, dull and uneven, with rough tufts where the hair had fallen out. His ribs stuck out under thin skin. His once proud neck had gone weak and frail.

Donna swallowed hard. A lump rose in her throat and made it difficult to breathe. Jack's one eye, which used to shine bright and alert, looked tired now, clouded with a quiet sadness that seemed to reach all the way into her soul.

He lifted his head slowly as they came near. A soft, raspy nicker escaped his throat. Once, that sound would have been full of warmth and welcome. Now it was fragile and desperate, the whisper of a horse begging for help.

"Oh, Jack," Donna whispered. Tears pricked her eyes. Her fingers itched to reach through the air and touch him, to stroke his coarse mane and promise him that everything would be all right. "What did they do to you, boy?"

Jay's deep voice cut through the heavy silence. He was not a man who asked permission when something was wrong.

"No one is going to starve these horses, Donna. Not on my watch."

His words were steady, filled with a quiet, unshakable resolve that somehow gave her strength. He was not giving the owner a choice. They were taking the horses, and that was that.

Donna looked over at him, grateful for his steady presence beside her. "I know, Papa." She used the nickname the grandkids had given him years ago. "But seeing him like this, it breaks my heart." Her throat tightened, and for a moment she could barely hold back the sobs.

Jay nodded grimly and reached over to unbuckle his seatbelt. "We're taking them home. Both of them. Jack and Cheyenne. That's final." There was no room in his voice for argument. Only resolve.

Donna glanced back at the paint mare. Cheyenne barely lifted her head, watching them with dull, tired eyes. She looked even worse than Jack, her skin stretched tight over her bones, her hips standing out like jagged hills. Her tail hung limp, and her eyes had gone faded and hopeless. Donna's heart clenched, a sharp ache twisting deep in her chest.

The memory of the old Jack came rushing back like a flood. The smooth, powerful strides he took when he ran free across a green pasture. The way he answered the smallest cue. His one eye sparkling with kindness. He was not just any horse. He was a horse with a story, and he deserved so much better than this.

"You remember the first time I rode him, Jay?" Donna murmured, her eyes never leaving Jack's fragile body. "He was like magic. You barely had to say a word. He listened, every single time."

Jay's lips twitched into a small, sad smile. "Yeah. That boy had more talent than most riders I've ever seen." His gaze drifted to the horizon, as if he were remembering it right alongside her.

Donna took a deep breath, drawing in the warm, dusty air. "The girl who rode him back then, she believed in him." Her voice was barely above a whisper, but it held a tiny spark of hope. "Maybe that's what he needs now. Somebody to believe in him again."

Jay reached out and unlatched the gate. The metal creaked softly in the quiet afternoon.

"Well," he said, "we're going to give him that chance. No matter what it takes." It was a quiet promise, a vow made to the horses and to Donna both.

The gate swung open.

Jack and Cheyenne stepped slowly into the paddock near the truck. Their hooves stirred up small clouds of dust that drifted in the warm breeze and settled like fine powder on the cracked earth. Jack moved with a careful, tentative grace, his head held low but his ears flicking toward them, as if he were trying to understand what was happening.

Donna felt the weight of years and heartbreak settle over her like a heavy cloak. The sun dipped lower, casting long shadows across the land and bathing everything in a soft amber glow. But beneath all of it, deep in her chest, a tiny spark of hope flickered stubbornly to life.

Maybe, just maybe, the horse nobody wanted could find a home again.

And with that thought, Donna smiled through her tears. The road ahead would be long, full of hard work and harder days. But Jack was not alone anymore.

Chapter 2

The Head That Came Down

A whole year went by at the 5 Daughters Ranch, and it changed everything for Jack.

The right food came twice a day, every day. A senior feed made special for older horses, with a supplement mixed in to help the two of them gain weight. There was coastal hay to nibble on and good green pasture to graze whenever they pleased. Donna brushed the dust from Jack's coat and the clouds from his eye, and slowly, the way the seasons turn, the old horse came back to life. His ribs disappeared under smooth muscle. His coat darkened and began to shine again. Cheyenne, the paint mare, filled out too, her white and brown patches bright once more.

And while Jack was growing strong, a little girl was growing up.

Bryer was two years old now. Her hair was long and light brown, streaked with gold from all her days in the sun, and it hung nearly to her waist. Most of the time it was twisted up into a knot, and most of the time that knot was working its way loose and falling into her face. Her skin was tanned a deep summer brown, dark from a life lived outdoors. And she had a fearless heart, with her grandmother Donna's spirit in her, the kind of spirit that runs toward things instead of away from them. She had her words early, too. By two she could say almost anything she wanted, and she usually did.

What Bryer wanted, more than anything, was to be in the pasture.

One bright morning, the sun stretched gold across the wide Texas sky. A gentle breeze carried the sweet smell of bluebonnets and fresh-turned dirt. Bryer burst out the farmhouse door, barefoot as always. Shoes, in her mind, were for indoors. The soft earth of the ranch felt like a blanket under her little feet.

"Bryer, wait up!" Donna called from the porch, half laughing, half worried. Jay followed behind, his heavy boots thudding against the dry ground.

Bryer did not wait. She trotted across the wide yard, straight toward the big black horse standing quietly by the fence. Jack. The one-eyed horse. The one everyone talked about and few dared to approach.

In the morning light, Jack looked like something out of a storybook. His black coat shimmered. His missing left eye gave him a mysterious, faraway look, as if he carried a secret behind that quiet, shadowed gaze. To grown-ups, he was a serious horse, well trained and careful, not to be rushed. But to Bryer, none of that mattered one bit.

She ran faster, her bare feet kicking up tiny puffs of dust. She did not slow down, not even when she reached his blind side. Most people thought that was dangerous, coming up on a horse where he could not see you. But Bryer did not know fear the way others did. At just two years old, with her long hair slipping loose from its knot and no shoes on her feet, she believed the whole world was full of friends waiting to be found.

Jack's ears flicked toward the soft sound of her feet. His one good eye found the little girl with the sun caught in her hair and no hesitation in her step.

"Bryer, be careful!" Donna's voice was sharp but gentle. Her hands were clenched at her sides. Jay's hands curled into fists too. A big horse, a tiny barefoot girl, and the blind side. It was a recipe for trouble.

But Jack did not move away.

Instead, something happened that none of them expected.

The big horse shifted his weight. The muscles under his coat softened. And then, slow and careful, Jack lowered his great head all the way down, down, down, until his soft muzzle was right in front of the little girl. He held it there, still as the pecan tree, as if he were saying, Here. I came down to you.

Bryer's whole face lit up.

She reached up with both tiny hands and laid them on his warm, velvet nose. She patted him. She giggled. She pressed her cheek against him and breathed him in, the sweet smell of hay and horse and morning.

"Hi, Jack," she said, her voice bubbling over with joy. "I like you. You're my horse now."

Jack let out a soft, low nicker. It was a tender, breathy sound, part greeting and part question, and only a few people had ever heard it from him. Those who had felt as if Jack were speaking straight to them in a secret language of trust.

Donna pressed a hand to her chest. Her eyes filled.

"Look at that," Jay breathed, his voice low with wonder. "He's never put his head down for anybody. Not like that. Not for anyone."

And that was the moment. Donna knew it the second she saw it. That was the moment a little barefoot girl fell in love with a one-eyed black horse, and the moment the horse decided she was his.

Jack kept his head low so Bryer could wrap her small arms around his muzzle. He was so gentle, so careful, as if he understood just how precious this fearless little girl was.

Donna stepped closer and knelt in the soft grass beside her granddaughter. "Bryer, honey, you're so brave," she said, her voice dropping to a tender whisper. "But Jack is a special horse. You have to be kind and patient with him."

Bryer lifted her little chin. "I'm kind," she said firmly. "And I'm not scared."

Jay chuckled, a deep, rumbling sound that filled the warm morning. "That's my girl. No fear in that heart."

The breeze stirred the smell of wildflowers and fresh hay, wrapping around them like a hug. Jack nuzzled Bryer's cheek again, his velvet nose warm against her skin. The big black horse with one eye and a hard past had found something new. A friend who saw him not as broken, but as whole.

For a long moment, the three of them stood there. The fearless little girl. The one-eyed black horse. And the two grown-ups watching with hearts full of hope. Overhead the wide Texas sky stretched on forever, bright blue and dotted with clouds.

From that day forward, Bryer and Jack were never far apart. Wherever she went, his soft nicker followed her like a song meant only for her. She ran barefoot through the tall grass, her laughter ringing out like a bell, and Jack walked close behind, his black coat gleaming in the sun.

Bryer taught everyone on the ranch that sometimes the bravest hearts come in the smallest packages. She showed them that courage is not about being big or strong. It is about believing in the good, even when other people do not. And Jack, the one-eyed black horse who had once seemed so distant, found a new life full of kindness and the soft, steady beat of little footsteps beside him.

Chapter 3

Learning to Trust Again

The sun was just peeking over the wide Texas horizon, painting the sky soft pink and gold. The air smelled of dew and wildflowers, and the morning was cool against Donna's skin as she walked out to the corral, her worn boots crunching on the gravel path.

Jack stood by the fence, and the sight of him still made her heart swell. His black coat shined now, deep and rich as polished stone. His ribs were gone, hidden under smooth muscle. His one eye was bright and clear, taking in the morning with calm, gentle curiosity. He stood tall and proud, as if he had never been anything but a champion.

Nearby, Cheyenne grazed in the soft light, her white and brown patches bright again, her tail swishing lazily at the morning flies.

Bryer was already outside, barefoot as usual, scrambling up the fence post with the fearless grace only a two-year-old could have. She leaned over and ran her small fingers through Jack's mane. "Good morning, Jack," she whispered. "You look so handsome today."

Jack answered with his soft, sweet nicker, the secret song that seemed to belong only to her.

Donna watched them and felt a quiet ache mixed with all that joy. Because Bryer would never know how it had really been.

She was just a baby that first year, far too little to remember. She would never remember how thin the horses had been the day they came home, how their bones had pushed against their skin, how Jack's once proud neck had hung low and weak. She would never remember the long, slow months it took to bring them back, or how many times the vet's truck rolled up the drive.

Getting a starving horse healthy again is not quick, and it is not easy. Horses do not fall that far and bounce right back. It takes the right food, given carefully and slowly so their bodies can handle it.

Donna and Jay fed them a senior feed made just for older horses, the kind that has everything an aging horse needs built right into it, and they added a supplement to help the two of them put on weight. There was coastal hay to eat, not the rich alfalfa hay that runs too hot for East Texas, and good green pasture to graze in any time they liked. And on the vet's good advice, Donna and Jay gave the horses a worming treatment to clear out all the parasites in their bellies. It was not just one dose, either. It was one dose every day for five days in a row, and then, sixty days later, they did the whole five day round all over again to be sure every last parasite was gone. It takes all of that, and it takes time, and the vet coming out again and again. And most of all, it takes love, the kind that shows up every single morning whether you feel like it or not.

For Jack, it was the steady kind of work. A little stronger each week. A little shinier. A little more like himself. There were worries along the way, and bumps in the road, but with patience and care, Jack climbed his way back.

For Cheyenne, it was harder. Much harder.

The old paint mare had been worse off than Jack from the very start. Donna and Jay gave her the same good senior feed, the same weight-gain supplement, the same gentle care. But Cheyenne just would not put on weight. She grew weaker instead of stronger, and Donna was truly afraid.

When the vet came out, he found the reason. Cheyenne had a deep infection inside her, a sickness that was making her too ill to gain weight no matter how much good food she ate. So the vet went to work on it. He tried one treatment. It did not work. He tried a second. It did not work either. He tried a third, and still poor Cheyenne could not turn the corner.

There was one Saturday Donna would never forget. After three treatments that had not worked, things looked just about as hopeless as they could look. With heavy hearts, the family had begun to wonder if the kindest thing might be to say goodbye to Cheyenne.

But that morning, Jay got a call. He was needed at work, and he had to go.

Donna always believed, deep down, that there was a reason for that call. Because with the goodbye put off, she found she could not stop fighting for that mare. She picked up the phone and called the vet right back out to the ranch.

The vet checked Cheyenne over again, slow and careful. Then he told Donna there was one last thing he could try, a treatment he had never used before. It was a long shot, a real last hope. "Let's give it one more week," he said. "We'll try this, and we'll keep working with her. If she starts to gain, we'll know she wants to fight. If she doesn't, then we'll know we did everything we possibly could."

One more week. One last hope.

Donna poured her whole heart into those seven days. And that long-shot treatment, the one the vet had never tried before, turned out to be the one that finally worked. Somewhere in the middle of that week, the smallest thing happened. Cheyenne began to eat.

Then she began to gain. A little at first, then a little more, until one morning Donna ran her hand along the mare's side and felt that the sharpest bones had softened.

Cheyenne had decided to live.

"You hung on, old girl," Donna had whispered into her mane that day, tears running down her face. "You hung on."

Now, watching both horses healthy and shining in the morning sun, Donna sent up a quiet thank you. Some things, she believed, were watched over by hands far bigger than her own.

Just then, Jay came out of the house, his boots crunching on the path, his calm smile making the whole ranch feel safe. Bryer hopped down from the fence and ran to wrap her arms around his sturdy legs.

"Papa," she said, bright and eager, "Jack and Cheyenne are all better now."

"That's right, little cowgirl," Jay said, ruffling her hair. "Good feed, good care, and a whole lot of love. That's what brought 'em back."

He looked at the horses a long moment, then crouched down to Bryer's level. "You know what, Bryer? These two are gonna be yours someday. Jack and Cheyenne both."

Bryer's eyes went wide as the morning sky. "Mine? Really?"

"Really," Jay said. "But you've gotta grow up first. Your mama, Krystal, she's scared of horses. She had a bad scare when she was young, and she won't have them at her place. So your horses are gonna live right here with us, at the ranch, where you can come love on 'em anytime you want."

Bryer thought about that hard, the way little ones do. Then she knelt in the warm dirt and pressed her cheek against Jack's neck. "I'll take good care of you," she whispered. "I promise."

Donna smiled, her eyes misty. "You're learning to trust again, little one," she said softly. "Just like Jack did."

Chapter 4

The Language Between Them

By afternoon, the Texas sun hung low and warm over the 5 Daughters Ranch. The light stretched long shadows across the dirt and turned the roads a rusty red. Bluebonnets and Indian paintbrush peeked from the fence line, bright against the scrubby brush, and the old windmill creaked lazily in the breeze.

Bryer's bare toes wriggled in the cool dirt as she toddled toward the black horse standing quietly under the old pecan tree. The shade there was a relief from the heat, a cool patch of ground where the air smelled of pecans and mossy roots.

Jack was no ordinary horse. His coat was as dark as midnight, so deep it seemed to drink up the sunlight. His body gleamed with health now, every muscle smooth and strong. But the most remarkable thing about him was his one eye, deep and dark, holding stories as old as the ranch itself. It was as if that single eye could see straight into your heart.

He stood still beneath the tree, his jet-black mane stirring in the breeze, his eye following Bryer's every move with calm, careful attention. To anyone else, Jack was a serious, well-trained horse, every muscle ready, like an instrument waiting for its cue. But to Bryer, he was something else entirely. A friend. A guardian. A quiet companion who understood her like no one else did.

Her long hair swung as she came close, the loose strands catching the sunlight in streaks of gold. She stopped just in front of Jack's broad neck and reached out a chubby hand.

Most horses do not like to be touched on the face. Jack especially. If a stranger reached toward the front of his head, he would pull away fast, ears pinned back, a clear warning to keep your distance.

But not with Bryer.

Slow and careful, Jack lowered his head until his warm muzzle rested right in her small hands. She scratched behind his ears. She ran her fingers gently along the side of his soft nose. And Jack stood as still and steady as the pecan tree itself, letting the little girl do what he let no one else do.

"Hi, Jack," she said in her small, sure voice. "You're my best horse."

Jack's nicker was soft, a gentle, breathy sound like a secret whispered just for her.

"You're not like other horses, Jack," she whispered, leaning her cheek against his neck. "You let me do stuff."

His eye glistened in the fading light, a slow blink full of trust. He watched her always, reading every little movement, every breath, taking in the language of her small hands and her gentle voice.

Bryer's mama, Krystal, had always been too afraid to come near Jack. His one eye, the way he sometimes startled, all of it made her nervous. But fearless little Bryer saw something in him that no one else did. Something gentle, something that had been waiting a long time for someone to understand.

Jack's "automatic buttons" were famous on the ranch. One touch of a rider's leg, one tilt of a rein, and he answered with the grace of a champion. But he was not a horse to take lightly. If you did not know what you were doing, he would push your buttons right back, and you would end up sitting in the dirt, wondering what just happened.

With Bryer, though, there was no need for reins or commands. Their language was quieter than that. It was the language of trust and patience, of small hands and soft voices, of secrets shared under a pecan tree.

Bryer reached for the low fence beside Jack, gripping the sun-warmed wood. She swung one leg carefully over the top rail and eased herself up beside him. Jack shifted his weight, calm and careful underneath her. Her little fingers found his mane, cool and silky in the shade, and held on, just enough to feel connected.

"Can you hear me, Jack?" she asked softly. "I'm gonna tell you a secret."

Jack's ears twitched forward, and he leaned his head a little closer, as if inviting her to share.

"I think you're magic," she said, eyes wide with wonder. "Like a big black knight who lost his other eye but still fights for the good guys."

Jack gave a low nicker, a sound full of quiet agreement. He did not need words to say he understood.

Up on the porch, Donna rested her hands on the railing and smiled as she watched the pair. The late light caught her silver hair and made it shine. "Those two have a language all their own," she said quietly to Jay, standing beside her.

Jay nodded, his eyes warm beneath his cowboy hat. "It's a quiet kind of talking," he said. "The kind that doesn't need words."

Bryer slid down from the fence and landed soft in the dirt. She patted Jack's shoulder like an old friend, and he lowered his head once more so she could reach the spot just behind his ear, the one place he let no one else touch.

"I'm the only one who gets to do that, huh?" she said, grinning.

Jack flicked his ears once, a tiny, almost invisible nod. Yes. Just you.

Then Bryer held out a handful of fresh green grass she had pulled from the pasture's edge. Jack sniffed it, and gently, so gently, took it from her fingers, careful never to nip.

"You're the best horse ever, Jack," she said, her voice full of love.

His nicker was low and soft, like a lullaby only she could hear.

Later, as the sun dipped behind the hills and painted the sky pink and orange, Bryer sat on the porch steps with Jack's heavy head resting beside her. The world was quiet, just the rustle of pecan leaves and the far-off call of a whippoorwill.

"No one else knows your secrets," she whispered. "Just me."

Chapter 5

Buttons and Bruises

The Texas sun stretched low across the wide, dusty corral of the 5 Daughters Ranch. The air was thick with the smell of dry grass and sun-baked dirt, and a soft breeze rustled the mesquite trees. Somewhere came the faint, steady clatter of horseshoes on hard ground.

Donna leaned against the weathered fence, the rough wood warm beneath her hands, her eyes full of quiet pride as she watched Jack. He was a striking sight now, his coat glossy and dark, shining even in the fading light. His one bright eye held a deep understanding, as if he could hear every word she whispered to him.

Jack flicked his tail at a few buzzing flies and let out a low, comfortable nicker. His muscles rippled under his shiny coat as he shifted his weight, his broad hooves stirring up small puffs of dust.

Donna smiled. Jack was no ordinary horse. He answered the lightest touch, a gentle press of a leg, a soft pull of a rein. She liked to call those little signals his "automatic buttons," because once you knew where they were and how to use them, Jack moved like magic.

But she also knew Jack had a stubborn streak. He could test a rider who did not understand him. He remembered every hurried grip and every wrong move, and he expected respect and patience in return. He was a horse of quiet dignity, never to be rushed.

Out in the corral, a splash of sunshine came bouncing barefoot across the dirt. It was Bryer, her long hair flying loose behind her, running like the wind. At two years old, she was as fearless as the Texas sky itself. She did not mind the heat or the grit beneath her bare feet. Her little toes dug into the dirt with every step, kicking up tiny clouds that glittered in the sun.

Her cheeks were flushed pink, her eyes shining with determination. She had made up her mind to say hello to Jack her own way. Bold and brave, just like always.

Without a moment's pause, Bryer took a deep breath and ran straight at Jack.

Now, horses like Jack have blind spots, places they simply cannot see, especially right under their noses and near their hooves. They depend on careful riders and slow, thoughtful movements to keep everyone safe.

Jack did not expect a whirlwind of a barefoot girl barreling straight at him. Startled, his strong body tensed, and he lifted a hoof without thinking. The edge of his hoof caught Bryer's big toe just as her little foot landed.

Bryer froze. Her eyes went wide with shock. Then a sharp cry burst out of her, loud as thunder rolling across the plains.

But Jack did not bolt, and he did not stomp down. Instead, he held perfectly still, his lifted hoof hovering just above the ground, careful not to press down even a little more. The soft, sandy dirt had cushioned the blow and kept it from being worse.

Donna's voice cut calmly through the noise. "Jack, pick it up."

Jack obeyed at once, lifting his hoof clear away, as if he understood every word. His one good eye looked gentle and worried. He gave a soft nicker, a sound full of apology.

Bryer's tears spilled down her cheeks, but even through the sting, she still wanted to be near him. "Feed him," she whimpered, reaching a trembling hand toward his muzzle.

Donna stepped forward and brushed a stray lock of hair from Bryer's sticky face. "Honey, you've gotta stop crying. You're scaring him. He doesn't understand why you're upset."

Bryer sniffled and pressed the back of her hand to her cheeks, wiping the tears away. Then she looked up at Donna, her eyes shining steady and brave.

Donna's smile was warm. "That's my girl."

Before Bryer could lose her courage, Donna scooped her up in strong, loving arms. Bryer felt the steady beat of her grandmother's heart and breathed in the familiar smell of soap and sun-warmed skin. Then, gentle as anything, Donna settled her onto Jack's broad back.

Jack nickered softly, as if to say, It's all right now. I'm here.

Bryer wrapped her small arms around his thick neck. She felt the warmth of him, the steady thump of his heart, and breathed in the sweet smell of hay and leather and earth. Jack stood calm and still, his one eye watching her with gentle kindness.

Together they walked slowly toward the feed pen, their shadows stretching long across the dusty ground, everything washed in honey-gold light.

As they went, Donna leaned close and whispered, "Jack's got automatic buttons, but he knows just when to be gentle. And you, Bryer, you've got a heart as big as Texas."

Bryer grinned, her toe still sore but her spirit shining brighter than ever. She felt proud and strong, riding her favorite horse like a real cowgirl.

On the 5 Daughters Ranch, every scrape and bruise came with a lesson. Lessons about kindness, about trust, and about the quiet kind of bravery that makes a girl and her horse the perfect pair.

Chapter 6

The Summer at A&M

The morning sun stretched golden across the wide Texas sky, warming the rolling fields of the 5 Daughters Ranch. A meadowlark sang somewhere nearby, its clear song weaving through the hum of cicadas hidden in the live oaks. But Donna was not thinking about the pretty morning. She was thinking about Jack's eye.

A few weeks back, Jack's one good eye had taken sick. It started small, a little red and weepy, but it grew worse instead of better. An infection had crept in, deep and stubborn. Donna and Jay had been treating it at home for more than six weeks, putting medicine in twice a day, morning and night. And still it would not heal.

Donna lay awake at night worrying. Jack had already lost one eye, long before he ever came to the ranch, most likely to an infection just like this one, back when whoever owned him did not go far enough to save it. This was the only eye he had left. The thought of him losing it, of Jack going dark and blind, made Donna's stomach twist.

She had already begun reading everything she could find about caring for a blind horse, because she would never give up on Jack, sight or no sight. But she was not about to lose his eye without a fight. Not when there was still something she could do.

And there was. There was one place that could give Jack a chance the ranch could not.

"We're taking him to A&M," Donna told Jay one evening. "The veterinary hospital. They can do more for that eye than we can do here."

Jay nodded slowly. He had been thinking the very same thing. "Then that's what we'll do."

So it was Donna and Jay who loaded Jack into the trailer that morning, and Donna and Jay who drove him the long road to the Texas A&M veterinary hospital themselves. Donna would not have it any other way. She walked him in, talked to the doctors, and stayed until she had seen him settled safe in his clean, quiet stall. Only then did she kiss his velvet nose and tell him to be brave.

"You be good now," she whispered. "We're gonna get that eye better. I promise."

At the hospital, the doctors went right to work. They ran a tiny tube up gently under Jack's eyelid, so they could send medicine straight to where the infection was hiding. And instead of twice a day, the way it had been at home, they treated that eye five times every single day. They would care for him like that for forty-five days, however long it took to win.

Donna's heart felt heavy as she drove away without him. But she knew, deep down, this was the best and bravest thing she could do for her old friend. "Hang on, Jack," she said softly. "Give it everything you've got."

There was one more thing about that big hospital that turned out to be a blessing. The whole building was cool and air-conditioned, top to bottom. Out in the open pasture, the fierce Texas summer sun could bleach Jack's black coat to a dull, faded brown. But inside those cool walls, day after day, out of the heat and the sun, his coat began to change. The faded brown slowly darkened, deeper and richer, back toward the jet black he had worn in his champion days.

Back at the ranch, a little girl missed him something fierce.

Bryer sat on the porch steps, her bare feet dusty and warm, her big brown eyes fixed on the empty paddock where Jack usually grazed. "Where's Jack, Grandma?" she asked, her voice small and worried.

Donna scooped her up and smoothed the hair back from her forehead. "Jack's getting some special help for his sore eye, honey," she said. "He's at a great big animal hospital, with lots of doctors who love horses and know just how to make him better."

Bryer twisted the hem of her sundress. "Will he miss us?"

"Oh, I'm sure he'll miss you just as much as you miss him," Donna said, pulling her close. "But you know Jack. He's got his own special way of making friends."

Bryer's eyes went wide. "How?"

"Well," Donna said, settling into the rocking chair with Bryer curled beside her. "Jack has the sweetest, most unusual nicker you ever heard. It's soft and warm, like a little breeze rustling through the trees on a summer evening." She paused, watching Bryer's face brighten. "Every time a doctor or a helper walks past his stall, he gives them that nicker. Just for them."

Bryer giggled. "Just for them?"

"Yes! And you know what happened?" Donna's eyes twinkled. "They all started carrying treats in their pockets. It got a little ridiculous."

Bryer's whole face lit up. "Like a secret club?"

"Exactly," Donna laughed. "Jack's got that whole hospital wrapped around his hoof. Every time somebody walks by, he gives that special nicker, and before you know it, he's got a pocketful of treats waiting for him. That sweet old horse is getting spoiled rotten."

Bryer sighed and rested her chin on Donna's arm. "I wish I could see him."

"Me too, Bryer. But he's in the very best hands. They're taking good care of his eye, and good care of the rest of him too."

Far from the dusty ranch, in a neat row of stalls at Texas A&M, Jack stood quietly in the cool, clean air. The smell here was different, a little medicinal, mixed with fresh straw. Five times a day, gentle hands tended his eye. And little by little, the redness faded. The infection began to lose its grip. The only eye Jack had left was healing at last.

Jack lifted that one good eye and watched the doctors and helpers move past his stall. With each one, he let out his soft, sweet nicker, a gentle call impossible to ignore.

One helper paused and smiled, reaching into her pocket. "Well, Jack, you sure know how to make friends, don't you?"

Jack nickered again, warm and grateful, and took the treat. Even the busiest doctor stopped a heartbeat longer, drawn in by the gentle black horse with one eye.

Back home, the summer days rolled on. Bryer chased butterflies and dug in the garden, but no matter how busy she got, there was always a little space in her heart saved just for Jack.

One evening, as the sky turned pink and lavender and the first stars blinked awake, Donna and Bryer rocked slowly on the porch.

"You know, Bryer," Donna said softly, "sometimes being apart for a little while makes the heart grow even stronger."

Bryer leaned against her shoulder. "Do you think Jack's thinking about us too?"

Donna smiled and brushed a strand of hair from her face. "I think he's thinking about all those treats he's getting. And maybe about the little girl who believes in him."

Chapter 7

Coming Home Different

The sun was beginning its slow, golden slide toward the horizon, painting the sky in warm gold and soft lavender. The air was heavy with the sweet smell of mesquite blossoms and the rich scent of earth from the past week's rain. It was that special kind of evening at the 5 Daughters Ranch when everything seemed to glow, as if the land itself was holding its breath, waiting for something wonderful.

Jack stepped carefully off the trailer, his hooves sinking into the soft, reddish dirt. His black coat shimmered in the fading light, smooth and flawless, without a single dusty patch or sun-bleached streak. Forty-five days inside the cool barn at A&M had kept the harsh sun off him, and now Jack looked simply magnificent.

Donna stood a few feet away, her eyes wide with admiration. She tucked a stray silver hair behind her ear, her hands trembling a little with excitement and relief. "Well, I'll be," she whispered. She brought a hand to her mouth as if to hold back a happy tear. "Look at you, Jack. You're like a shiny black stallion right out of a storybook."

Jay leaned against the fence post, arms crossed, a slow smile spreading across his sun-weathered face. "Never seen a coat that dark," he said, his voice deep and steady. "We could put you in the advertisements for good grooming, that's for sure."

Emma, one of the older granddaughters, moved closer, circling Jack with the easy grace of someone who had grown up around horses. Her dark brown hair hung loose down her back, and her skin was tanned deep from a summer spent outdoors. She ran a hand along his smooth, powerful muscles. "He looks like he belongs in a show ring again," she said, her face bright. "Like he's ready to win ribbons."

But all the grown-ups' fancy words did not hold a candle to what came next.

Barefoot and wild as ever, Bryer came barreling across the yard, her little feet kicking up clouds of dust. The cool grass brushed her ankles, but she did not slow down for a second. Her long hair, half loose from its knot, swung around her flushed cheeks. She did not even glance at the shiny coat or the way Jack's mane caught the last of the sun. She ran straight for him with her arms wide open.

"Jack! Jack, I missed you!" she called, the words tumbling over each other like a rushing creek.

Jack lifted his head slowly, his one bright eye catching the light. A soft, sweet nicker slipped from his throat, the one only Bryer truly understood. It was a sound full of welcome and belonging, a melody that said, I missed you too.

Bryer leapt forward and wrapped her small arms tightly around his neck. Jack lowered his great head so she could rest her cheek against his warm muzzle. His whiskered chin tickled her skin, and she giggled, the sound bright as chimes in the summer breeze. The bond between them was whole again in an instant, as if the weeks apart had been nothing but a blink.

Donna watched, her heart so full her eyes pricked with tears. "That's love right there," she said quietly to Jay. "Doesn't matter how shiny or new he looks. To Bryer, he's just Jack. The same friend she's always known."

Jay smiled. "You're right, Donna. Some bonds don't change, no matter how much time goes by." He shifted his weight against the fence. "And I reckon that black, one-eyed horse has a heart as big as Texas itself."

Krystal stood a little behind the group, her arms folded tight across her chest. She still felt a knot in her stomach when she looked at the big horse, old fears stirring in her mind. But seeing Bryer so fearless and happy, wrapped around Jack like he was the safest place in the world, made her shoulders soften just a bit. Maybe, she thought quietly, maybe Jack was truly home now.

Jack shifted his weight and gave another gentle nicker, nuzzling Bryer's hair with his soft muzzle. She reached up to scratch behind his ear, just like always. It was the same dance, the same friendship, the same quiet understanding that had never faded.

The sun dipped lower, the colors deepening to rich orange and purple. Long shadows stretched across the ranch, and the first fireflies began to blink, tiny lights flickering like stars come down to earth.

Jack, the one-eyed black horse with his glossy coat, was home. Not just in the dirt and grass and wide Texas sky, but right where he belonged, in the heart of a little barefoot girl who loved him exactly the way he was.

Bryer pulled back just enough to look into his one shining eye. "I'm so glad you're back, Jack," she whispered, stroking his cheek.

Chapter 8

The Horse Who Kept Her Secrets

Now that Jack was home and shining and well, the very best days began. Bryer had turned three by then, taller and busier and full of more words than ever, but just as barefoot and just as fearless. These were long, golden, ordinary days, the kind a little girl never forgets.

Bryer spent every minute she could out in the pasture with him. She did not need games or toys. She just needed Jack.

Feeding him was her favorite job of all, and she took it very seriously. Donna would scoop the feed into the bucket, and then Bryer would carry that bucket all by herself, hauling it across the ground with all her might. It was a heavy bucket for such a little girl, but Bryer never complained, not even once. She just set her jaw and lugged it along, independent as could be. Together she and Donna would open the gate and walk into the feed pen side by side. Then Bryer would stretch up on her toes and try her very hardest to hang the bucket on the panel, and Donna would help her with the last of it, because in truth that bucket was just too heavy for a tiny girl to lift up that high.

And here is the wonderful thing about Jack. He stood back and waited, patient as a Sunday morning. He did not so much as dip his nose toward that bucket until every last hand was off of it.

That was not a manner Jack came to the ranch with. Donna had taught it to him, and she had insisted on it, because a big horse shoving his heavy head into the bucket while you are trying to hang it is just plain rude. Jack had lived a hard life before he ever came to the 5 Daughters Ranch, but he was a smart old horse, and he was willing to learn. He learned to wait. He learned his manners. And only when the bucket was hung and all the hands were clear would he finally step up, drop his nose in, and begin to eat, his soft lips sometimes brushing Bryer's palm like a quiet thank you.

She would brush the dust off his jet black coat while he ate. And most of all, she would talk to him.

Oh, how Bryer talked to that horse.

Sometimes she would plop right down in the grass beside him and talk while he grazed. Bryer was seldom still, so even sitting she kept her hands busy, picking at the blades of grass while her mouth ran a mile a minute. That girl had an opinion about everything, in the very best way. She was sure of herself, certain of exactly who she was, the same way her mama Krystal had been even as a little bitty thing. If Bryer wanted to crack a joke, she cracked it. If she was happy, you knew it. If she was put out about something, you knew that too. Whatever Bryer was feeling, that is just what you got, and Jack got all of it.

And almost always, she came to him barefoot. If she had dressed herself that morning, her clothes were likely mismatched, and more than once she came running out to the pasture still in her pajamas, never thinking a thing about it. Out here, that suited everybody fine. There was no one to impress, only a one-eyed horse who loved her exactly as she was.

Other times, when the secrets were the important kind, the grass was not close enough. Jack was much, much taller than she was, so Bryer figured out a way to get them face to face. She would climb up the wooden fence, rung by rung, until she was high enough to look right into his one good eye. Then Jack would step in close, and the two of them would stand there nose to nose, the little girl and the great black horse, having their talks.

Nobody ever knew what she told him. Not Nana, not Papa, not her mama or her daddy. To this very day, no one knows Bryer's secrets, because the only one she told was Jack, and Jack never breathed a word.

That is the wonderful thing about horses, you see. A horse keeps your secrets. You can tell a horse anything at all, the happy things and the sad things and the maddest, most grumbly things, and a horse will never tell another living soul. A horse will never repeat it. A horse will never tease you about it or carry it off to somebody else. Whatever you give a horse to hold, a horse holds it forever. That is one of the very best reasons in the whole world to love them. Bryer's Nana knew that better than most, because she had been telling her own secrets to horses since she was a young girl herself, and not one horse had ever let her down.

And little Bryer had plenty to tell.

Sometimes she was mad at her big sister, or at one of her brothers. Up the fence she would go, leaning in close to whisper it all to Jack. He would flick an ear and listen, and somehow, by the time she climbed down, she did not feel quite so mad anymore.

Sometimes she was mad at her mama, or her daddy, or even at Nana and Papa. Especially when the answer to something was no. Then off she would march to find Jack, and Jack Jack, as she called him when it was just the two of them, would lower his big head and listen to every word.

But more than anybody, Bryer got mad at the cows. The cows were always doing something to ruffle her feathers. Some of them liked to head butt, lowering their broad heads to bump and shove, and that made Bryer madder than a wet hen. Sometimes she would hold out a treat just so, and the cow would not take it the right way at all. And every now and then a cow would take its treat wrong and bump her little hand with its mouth. Now, a cow only has bottom teeth in the front, not top ones, but to a three-year-old, a slobbery cow nibble was a terrible insult indeed.

When the cows wronged her, there was only one thing to do. March straight to the pasture, climb the fence, and tell Jack all about it.

"You won't believe what those cows did, Jack Jack," she would huff, her face screwed up tight. And Jack would listen, calm and patient, as if he agreed that the cows had behaved very badly.

Now, there was one more thing about Bryer that everyone on the 5 Daughters Ranch knew. When she got told no, she had a habit of going right ahead anyway.

"Bryer, you cannot go down to the cows by yourself," Donna would say.

"Okay, Nana," Bryer would answer, sweet as pie.

And the next thing Donna knew, that barefoot little girl had slipped out the door and was headed down toward the pasture, on a mission all her own. Bryer was famous for it. Tell her no, and somehow she would still end up outside, finding her way to her horse. It worried Donna and Jay to no end, a tiny girl wandering off toward the big animals. But it also told them something true about Bryer's heart. When she loved something, nothing on earth could keep her from it.

And what she loved was Jack.

She would find him out in the green grass or call him over to the fence, climb up to meet his eye, and unload her whole busy day. The good and the bad, the happy and the mad. Jack took it all. He never once told her to hush. He never said she was being silly. He just stood there, steady and warm, soaking up her words like sunshine, keeping every single secret safe.

"You're the only one who really listens, Jack Jack," she told him one evening, leaning her forehead against his.

Jack gave the softest nicker, that secret song meant only for her. And maybe that was his way of saying, *Your secrets are safe with me. They always will be.*

The sun sank low and golden over the pasture, painting the grass in honey light. A little barefoot girl and a great black horse stood face to face at the fence, sharing the quiet, sharing the secrets, two friends who understood each other better than anyone else ever could.

Chapter 9

Bryer's Turn (and Everybody Else's)

If there was one thing on the whole 5 Daughters Ranch that Bryer did not like to share, it was riding time.

In her mind, the horses were hers. All of them. Jack most of all, of course, but Cheyenne too, and even big, gentle Bailey. As far as Bryer was concerned, the only person who ought to ever sit on a horse was Bryer.

Now, Bryer was the only one small enough to be set right up on Jack's or Cheyenne's back the easy way. She hardly weighed more than a feed sack, so the grown-ups could slip a halter on the old horses and lead her around the pasture at a slow, gentle walk, no saddle needed. She would sit up there like a queen on her throne, beaming down at the whole wide world.

But Bailey, he was the riding horse for everybody else. Aunt Casie Ann would saddle him up, slide his bridle on, and lead the other kids around for their turns. And that, in Bryer's opinion, was a problem.

Because the other kids wanted to ride too.

Mostly it was Bryer and her sister Emma who loved it best. But every now and then Briggs, Jase, and Kohen would decide it was their turn, and here they would come. And on the big days, the family get-together days, when the whole crew showed up, why, there were riders everywhere. Sissy, who loved to ride as much as any of them, and the cousins RaeLynn and BraeLynn and SkyeLann, all of them lined up and waiting for a turn on Bailey.

This did not sit well with Bryer. Not one bit.

She would plant her bare feet in the dirt, fold her little arms, and scrunch up her face like a thundercloud. All these people, on her horses? It was almost more than a three-year-old heart could bear.

But the grown-ups had a secret weapon. Two of them, actually. Cookies and cows.

When it was somebody else's turn, Aunt Casie Ann would scoop Bryer up and say, "Come on, Bryer. Let's go see the cows." And off they would go, down to the fence with a whole stash of treats. Because if there was one thing Bryer loved nearly as much as Jack, it was bossing those cows around with a snack in her hand.

You had to come prepared, though. The cows here were particular, and so was Bryer. You needed vanilla wafers. You needed oatmeal pies. Oreos were a must, and graham crackers, and a few donuts too if anybody had them on hand. With a proper supply of cookies, Bryer could stand at that fence feeding the cows for ages, perfectly content, while everybody else got their time on Bailey.

And those cows did love their treats. They would crowd up to the fence, big and slobbery and eager, and Bryer would hand the cookies out one at a time, telling each cow exactly how to behave. It worked like a charm. Most of the time.

The family days were the best days of all. They came around regular, the whole bunch of them filling up the ranch with noise and laughter. There were too many cousins and aunts and uncles to count, Morgan and Sissy and Candice and Krystal, Aunt Casie Ann, and a long line of little ones from the biggest down to the smallest.

And there was always, always something to do.

Some of the kids went down to the stock tank to fish. It was stocked full of bass and perch, just right for little fishing poles and big proud smiles. Some of them built forts out of whatever they could drag together. Some played games out in the grass until they were red-cheeked and grinning. Nobody sat around staring at a screen. Out here, there was a whole world to run wild in, and the grandkids used every inch of it.

A few of the grandkids even liked the work best of all. Briggs and Jase would come looking for chores the minute they arrived, because chores meant they could earn a little money. Pawpaw had it all planned out. When the boys came to stay, he lined up enough jobs that a hardworking kid could earn a whole week's wages before it was time to go home. Emma liked earning her keep too. But Pawpaw always made sure there was plenty of fun mixed in with the work, because a ranch should never be all chores and no play.

Through all the splashing and fishing and fort-building and chore-doing, Bryer kept one eye on the horses at all times. And the moment a turn came open, she was right there, ready to climb up on Jack and be led around the pasture, the happiest little cowgirl in all of Texas.

Chapter 10

When Running Hurts

Time has a way of slipping by on a ranch, quiet as a breeze through the mesquite. The golden days rolled on, one into the next, and Bryer stayed right by Jack's side through every one of them. But even as the little girl grew, Jack was growing older.

He stood in the soft morning sun, his black coat shining like polished stone in the golden light. The air smelled of wild sage and damp soil from last night's rain. But something was different today.

Donna sat on the weathered porch and watched him carefully. Her eyes traced the familiar, proud shape of the one-eyed horse she had loved for so long. And she saw something new. A weariness. Jack was not as spry as he used to be. He shifted his weight slowly from one leg to the other, his movements stiff and careful, like an old clock ticking a little slower than before.

Donna's heart tightened. "Old age is catching up with you, boy," she whispered, her voice barely louder than the wind.

Jack had always been a proud and noble horse, trained at the highest levels on fancy ranches far away. His automatic buttons had always been perfect, a secret language between horse and rider. But now, even the smallest step seemed to hurt him. His strong legs, once like pillars, moved as if they carried a heavy, invisible weight.

Little feet pattered across the warm porch. Bryer came toddling out, her long hair swinging loose. "Hi, Jack!" she called, bright and fearless, and ran to him.

She reached up to stroke his muzzle, her small hand warm against his soft nose. Jack nickered, but today the sound was quieter, almost a whisper carried off by the breeze. Bryer's brow furrowed.

"Jack's slow," she said, puzzled, watching him try to lift a heavy leg and wobble before setting it back down. She had never seen him move like this. Usually he was fast and strong, like a big shadow racing across the pasture. Now he looked tired, as if his legs were made of heavy stones.

Donna knelt beside her granddaughter. "Jack's getting old, sweetheart," she said gently, tucking a loose strand of hair behind Bryer's ear. "Sometimes, when horses get older, they get something called arthritis. It's a little like when your knees get sore after you play too long. It makes their joints swell up and hurt."

Bryer tilted her head. "Joints?"

"Yes," Donna said. "Joints are the places where your bones meet, like your elbows and your knees. When they get swollen, it's hard to move. Jack's joints are getting stiff and sore, so running and playing hurt him now."

Bryer looked back at Jack, who was trying to step over a low branch on the ground. He stopped halfway, his hoof hovering uncertainly. The morning light caught the swelling along his legs.

"He can't run like before," she said softly, a little sadness in her voice.

"No, honey," Donna said. "Jack's still here, but he needs to rest more now. We have to be patient and give him lots of love."

Over the next few weeks, the whole family noticed the changes, little by little. Jay stood often by the fence near the barn, watching Jack move slowly through the pasture. "He's been a champion," Jay said one evening, his voice steady but laced with quiet sadness as the sun dipped low. "That old horse has given us a lifetime of memories."

Emma stood nearby, her arms crossed. "But why does he look so sore all the time?" she asked, full of worry.

Donna sighed. "It started out as arthritis, Emma. That's when the joints get sore and swollen. But Jack's has turned into something harder. The vet has a name for it. Degenerative joint disease. That's a big set of words, but all it means is that his joints are wearing out, and they keep getting worse no matter what we do. His ankles and his knees just keep swelling up." She rested a hand on Emma's shoulder. "We're doing everything we can. We just have to be patient with him now, and love him through it."

Bryer still visited Jack every single day, her bare feet padding across the warm dirt. She would sit beside him under the wide Texas sky and whisper stories to him, tales of adventures and magical places, and sing little songs in her soft, sweet voice. Jack would nicker back quietly and lean his heavy head gently against her small shoulder.

One late afternoon, when the sky was painted purple and gold, Donna found Bryer sitting by Jack's side, tears glistening in her big eyes.

"Jack's hurting, isn't he?" Bryer asked, her voice trembling.

"Yes, baby," Donna said, wrapping her arms around the little girl. "He is. Sometimes even the strongest horses get tired and need a lot of help."

Bryer looked up, searching her grandmother's face. "Will he be okay?"

Donna's heart ached, but she smiled softly. "We'll take care of him, always," she promised. "But we have to watch him close and make sure he's not hurting too much."

As the weeks went by, Jack's joints grew worse instead of better. His ankles and his knees swelled bigger and bigger, swelling that no medicine could bring down. The kind country vet came out again and again. He tried everything he knew. But each time, he shook his head a little more sadly. There was no medicine that could fix what was happening, he told Donna gently, and nothing that could take the pain away for good. Jack's joints were swelling faster than anyone could control.

Donna and Jay had given that horse every bit of love and care and medicine there was. They had fought for him for years, the same way they fought for him the very first day. They had never once given up on him. But deep down, they knew what they did not want to know. There comes a time when the kindest thing you can do for a hurting friend, when there is truly nothing left to make him better, is to let him go gently, with love.

The days grew quieter for Jack. The wind still whispered through the mesquite, and the sun still warmed his black coat. But the running, the jumping, the fast games, they were slowing down, just like Jack.

One evening, Donna reached out and touched Jack's soft nose as he came to the fence. His breathing was slow and heavy under the silver moonlight. "We'll face this together, old friend," she whispered. "You're not alone."

Nearby, Bryer listened with wide eyes, her small hands gripping the edge of the porch. "Will Jack still be my friend if he can't run?"

Donna smiled and brushed a strand of hair from her forehead. "Always, sweetheart. Some friendships don't need running. They just need love."

Chapter 11

The Last Bath

The sun was just stretching its golden fingers across the wide Texas sky, painting the horizon soft pink and orange. The air was cool and fresh, carrying the faint scent of sagebrush and wildflowers across the 5 Daughters Ranch.

The night before, Donna and Jay had sat the girls down and told them the truth, gently and quietly, the way you hand a child something fragile. Jack's old body had grown too tired and too sore. His joints hurt him with every step now, and there was no medicine left that could take that pain away. The kindest, most loving thing the family could do was help him go on to heaven, where he would be young and strong again and never hurt for even a moment.

It was a hard truth, but Donna believed with her whole heart that children can hold hard truths when they are wrapped in love. And she believed something else, too: that this was not the end of Jack. It was a beginning. He was going to meet God.

So in the morning, the girls knew exactly what they were doing. They were not just giving Jack a bath. They were getting him ready. Ready to go to heaven. And they were going to send him off shining.

"Come on, girls," Donna called softly from the porch of the big red barn. "It's time to get Jack ready."

Bryer ran barefoot across the porch, and Emma followed close behind, already working the tangles out of Bryer's long hair and smoothing it back into its knot. There was a hush over the morning, but it was not a sad hush. It was a careful, tender one, the kind that settles over you when you are doing something that matters.

Jack stood quietly in the shade of the open barn doorway, his black coat catching the first light. Though he had only one eye, it was bright and full of wisdom, calm and steady as the children came close. He gave his sweet, low nicker, the one that always seemed to say, I'm here. I trust you.

Bryer reached up and laid her small hands along the smooth side of his neck. "We're gonna make you so pretty, Jack," she whispered. "Because today's a special day."

Emma nodded, blinking fast but lifting her chin. "We're gonna make you shine for heaven."

Donna's chest went tight, but she held herself steady. She picked up her camera. Today is for them, she told herself. Let them have their joy.

They led Jack gently to the wash rack beside the barn, where the hose hung ready. When Emma turned the nozzle, a soft stream of water sent tiny droplets sparkling in the sunlight like a shower of diamonds.

Bryer was far too short to reach Jack's broad back on her own. She was only three, and Jack was a tall, tall horse. So she did what she always did at bath time. She flipped over her sturdy bucket, the same one she used every time, and climbed up to stand on top of it. Even then she had to stretch, up on her toes, to pour the water slowly over his back. It soaked into his coat, darkening it like wet velvet.

Jack did not flinch. He stood still as a statue, his tail swishing now and then, soaking up every bit of the attention. He loved it. You could see it in the soft, half-closed look of his one good eye.

Emma squeezed a dollop of shampoo into her palm, sweet-smelling, like wildflowers and honey, and worked it into his mane. Bryer joined in, her tiny hands brushing the suds through his coat. The bubbles popped against their skin and made them giggle, and soon the two girls were laughing, splashing, scrubbing every inch of the big black horse who stood there happy as could be, eating up all the love.

Through the camera, Donna watched them and smiled. She caught Bryer's grin, mid-laugh, a smear of suds on her cheek. She caught Emma braiding the wet mane into neat plaits that shone like ribbons of silk. She caught Jack's contented eye and his swishing tail. Click after click, she gathered up these last bright moments to keep forever.

She was doing just fine. She was smiling. She was steady.

And then Bryer climbed up on the fence.

The little girl perched on the weathered rail, leaned toward Jack, and began to tell him about heaven.

"You're gonna love it, Jack," Bryer said, her voice soft and sure. "It's the most perfect place there is. God's gonna fix your legs so they don't hurt anymore. Not even a little."

Donna's hands went still on the camera.

"There's gonna be green grass everywhere," Bryer went on, eyes shining. "So much green grass you can't even see the dirt. And you can run, Jack. You can run as fast as you want, all day long, and you won't ever get tired and you won't ever hurt again."

That was the moment.

That was when the first tear slipped down Donna's cheek, warm and quick. She wiped it away fast, before either of the girls could see, and steadied the camera again. Because this moment did not belong to her sorrow. It belonged to Bryer and Emma and the brave, beautiful faith pouring out of a three-year-old's heart. It belonged to Jack, listening with his ears tipped forward as if he believed every single word.

"You won't be scared," Bryer told him gently. "And you won't be alone. God's gonna take real good care of you. And someday, I'm gonna see you again."

Jack lowered his head toward her and gave the softest nicker, that secret song that was only ever hers. Bryer giggled and pressed her forehead to his.

Emma came and stood beside her sister, resting a hand on Jack's wet shoulder. "We love you, Jack," she said quietly.

Donna lowered the camera at last and looked at the three of them in the golden morning light. Her heart was breaking and full at the very same time. The girls were not afraid. They were not falling apart. They were loving Jack all the way to the gate of heaven, sending him off clean and braided and shining, the way you send off someone you love when you know, deep down, that goodbye is not forever.

Jay came up beside her and rested his broad hand on her shoulder. They shared a look that said everything without a word. The kind country vet would be here soon. But this part, this last bath, was exactly the way it should be.

"They're doing real good," Jay said low.

"They are," Donna whispered. "They understand better than most grown folks do."

Chapter 12

Heaven's Pasture

That evening, the sun dipped low over the wide Texas sky, casting a warm, golden glow across the rolling fields of the 5 Daughters Ranch. Puffy clouds drifted overhead, tinged pink and gold, as if they too were settling in for the night. A gentle breeze carried the sweet smell of mesquite blossoms, and somewhere a cricket began its soft, steady song.

Jack stood clean and shining out in the pasture, his mane still braided from his morning bath, his coat gleaming like polished stone. He had spent the whole day wrapped up in love, soaking in every splash and giggle and gentle hand. Now the day was winding down, and the ranch was growing quiet.

It was time for the girls to go home.

Donna and Jay had thought long and hard about this part. They had been honest with the children that Jack was going to heaven, and they had let them send him off that morning with joy, getting him braided and shining and ready. But the very last moment, the quiet goodbye at the end, was not a thing for little ones to carry. Some kindnesses are done softly, out of sight, so that a child's heart can hold onto the good and the beautiful. That was a weight for grown-up shoulders.

So in the golden evening light, Bryer and Emma hugged Jack one more time.

Bryer reached up and wrapped her arms around his big neck, breathing in the clean, sweet smell of him. "Bye, Jack," she whispered. "I love you."

Jack lowered his head and gave the softest nicker, that secret song meant only for her.

Then Krystal loaded the girls into the truck and they headed off down the long dirt road toward home, little hands waving out the window until the dust rose up behind them and the ranch slipped out of sight.

The ranch was very quiet that night without them.

The kind country vet came the next morning, when the sun was warm and golden over the pasture.

Helping a beloved animal leave the world gently, when his body has grown too tired and too sore to carry him without pain, is one of the hardest and most loving things a family can ever do. The vet was quiet and gentle. First he gave Jack a shot to take away the soreness, the deep ache that had settled into the old horse's joints. And for the first time in a long while, Jack felt no pain at all.

He lay down in the soft grass, calm and relaxed, his breathing slow and easy. He looked, for all the world, like a horse simply settling in to rest in the warm morning sun.

Donna knelt beside him in the soft grass. There was one more thing she wanted to give him. One last gift.

She pulled out her phone and called Bryer's daddy, Brent. He answered, and Donna asked him to turn on the camera so the two of them could see each other. In a moment, there was Bryer's face, glowing on the little screen, smiling that bright, fearless smile of hers.

Donna held the phone down low, close to Jack, so he could see it. So he could see her.

"Look, Jack," Donna said softly. "Look who came to see you."

Jack's one good eye found the screen and settled there. And on it was Bryer, her long hair slipping loose the way it always did, her face lit up with love.

"Hi, Jack," Bryer said, waving her little hand. To her, he was just lying down in the grass, resting and happy, soaking up the warm morning sun the way horses do. "I love you, Jack. You're the best horse ever."

Jack gazed at her sweet face, glowing on the little screen, and gave one last quiet breath of a nicker. The last thing the old one-eyed horse ever saw in this world was the face of the little girl who had loved him most, the girl who had always believed in him.

And then, peaceful and unafraid, with no more pain to carry, Jack closed his eye and slipped away, off to that green and endless pasture Bryer had told him about, where the grass goes on forever and a horse can run and run and never, ever hurt again.

Donna stayed there a long while in the quiet of the morning, her hand resting on his warm neck, the sun spilling gold across the pasture. Tears ran warm down her cheeks. Her heart was heavy with the weight of goodbye, but it was full too, full with the comfort of knowing Jack was free now. Whole and strong and young again. Running where the grass never ends.

She thought about Bryer's face on that little screen, the last thing Jack ever saw, and she knew she had given him the most precious gift she had to give. Not a fancy goodbye. Just love. The same love that had carried him all the way from a bare and dusty pen to this soft Texas grass under a sky full of stars.

"You were so loved, old friend," she whispered. "All the way home."

Chapter 13

What Jack Left Behind

The sun was dipping low over the wide Texas horizon, painting the sky in shades of melted gold and soft apricot. The warm light stretched across the dusty paddocks of the 5 Daughters Ranch, casting long shadows that swayed in the evening breeze.

Donna stood by the weathered fence, her fingers tracing the rough, sun-warmed rails, her eyes resting on the empty spot where Jack used to stand. The one-eyed black horse had been such a striking figure on this open land. Now that space was quiet, but his presence still lingered, soft as the smell of mesquite blossoms on a summer night.

Donna's heart ached with a tender, bittersweet sorrow. She could almost feel Jack's warm breath, almost hear the gentle nicker he gave whenever he saw Bryer coming. And her thoughts turned to Bryer, her wild-haired, barefoot granddaughter who had loved

Jack with a heart as wide as the Texas sky.

Bryer was just three years old, barely tall enough to reach the saddle horn, her skin tanned deep brown and her long, gold-streaked hair catching the wind like a banner.

Donna sighed softly. "If only," she whispered to the fading light. If only Bryer and Jack could have grown up together, could have ridden into the big shows side by side, ribbons fluttering in the hot Texas sun. But time does not work that way. Jack had already lived a long life by the time Bryer came along, and his racing days were behind him. And Bryer was still far too little to guide him through the buckles and ribbons of competition.

Jack had been a legend around these parts. He could change direction with a brush of a leg, halt at a whispered word, and spin on a dime, but only for someone who truly knew what they were doing. If you did not, well, Jack had his own way of teaching you, until you either rode right or found yourself sitting in the dust.

Donna's lips curved into a small smile as she remembered his unique, sweet nicker, the sound that had always made Bryer giggle. It was a secret song, a language all its own, spoken only between the two of them. Watching Bryer and Jack together had been like watching two old souls in a quiet conversation, full of understanding and love.

There were no ribbons or buckles hanging in the barn for Bryer and Jack. That was the bittersweet truth Donna held close.

How she wished, just once, that the two of them could have gone into an arena together. Donna had no doubt at all how it would have gone. With Bryer's fierce determination and all the skill Jack still carried deep in his bones, the two of them would have won championship after championship. But it never would have been about the trophies. Not for Donna, and not for those two. It would have been about watching them out there in the ring, moving as one single thing, the way they already did everywhere else on the ranch.

Because that was the rarest part of all. Bryer and Jack did not really speak in words. They spoke heart to heart, and mind to mind. Anyone lucky enough to have once had a true heart horse knows exactly what that means. It is a closeness deeper than any command, deeper than reins or cues, a quiet understanding that passes between a person and an animal who simply belong to each other.

That was Bryer and Jack. And the saddest, sweetest truth of their whole story was this: they had found each other too early in her life, and too late in his. Jack had grown too old to compete by the time Bryer was big enough to ride, and Bryer was far too little when Jack was still in his prime. The arena, the ribbons, the buckles, all of it stayed a dream, because the years had simply not lined up the way a heart would have wanted them to.

But Donna's gaze softened as she watched Bryer now, reaching out to stroke the rough wood of the fence, as if trying to touch Jack's spirit in the warm, fading light. "Jack gave her something no buckle ever could," Donna murmured. "Love mattered more than any ribbon."

What Jack left behind was so much more than skill or trophies. It was something deeper, something you could not hang on a wall. It was the quiet patience of waiting for a signal. It was the courage to be fearless even when the world feels too big. It was the trust that grows in the softest moments shared between a girl and her horse.

Bryer carried those lessons with her like treasure. Even now, she spoke the language of horses without words. She knew when to wait and when to move. She knew how to listen to the rhythm of hooves and the whisper of the wind. Her hands were steady and sure, and her heart was wide open.

Donna thought back to all those afternoons on the porch, the air sweet with honeysuckle, watching Bryer and Jack together. Bryer's laughter ringing out like little silver bells. Jack's soft nicker answering back, a lullaby just for her. Those sounds had woven together into a song of love and trust she would never forget.

The sky deepened to dusky lavender, and the first stars began to twinkle, delicate pinpricks of light across the endless Texas canvas. Donna felt a warm hope settle in her chest. Jack's spirit was alive in Bryer. Wild and free. Steady and strong. The ribbons and buckles might never come, but that did not matter at all. Because Jack had given Bryer the greatest gift of all: a lifetime of love, and the courage to believe in herself.

As Bryer turned toward the house, her bare feet kicking up little clouds of dust, Donna watched her with gentle pride. The little girl's cheeks were flushed from the day's adventures, her skin glowing in the last amber light. Donna knew, without a single doubt, that Bryer would always be a horse girl. Not because of trophies or shows, but because of the heart Jack had left beating in her own.

And that was more than enough.

About the Author

Donna Minton Williams is a wife, a mother of five daughters, a grandmother of thirteen, and a great-grandmother of one. She and her husband, Jay, run the 5 Daughters Ranch in LaRue, Texas, where they raise registered show cattle and welcome a houseful of barefoot grandchildren who fill the place with laughter.

Donna has spent a lifetime with horses, beginning when she was just eighteen years old. She believes with her whole heart that animals can heal us, teach us, and keep our secrets, and that no child should ever be turned away from the chance to know that kind of love.

The 5 Daughters Ranch series grew out of the real animals and real children Donna loves. Jack and the Girl Who Believed in Him is the true story of her granddaughter Bryer and a one-eyed black horse who finally found his way home.

Donna writes from her faith and her family, and she hopes these books remind every child of something she holds dear: the day you stop dreaming is the day you stop living.

A Guide for Teachers and Families

Jack and the Girl Who Believed in Him supports reading comprehension, vocabulary, and social and emotional learning for readers ages 8 to 11. These prompts work in the classroom or at home.

QUESTIONS TO TALK ABOUT

1. When Jay first sees Jack and Cheyenne starving, he decides on the spot that both horses are coming home. What does that choice show about how Jay acts when something is wrong?
2. Bryer runs up to Jack on his blind side, where a one-eyed horse cannot see. Why do you think Jack lowered his great head to her instead of pulling away?
3. Cheyenne almost did not make it, but Donna called the vet back for one more week and one last try. What does this teach about not giving up too soon?
4. Nana believes a child can hold a hard truth when it is wrapped in love. How did the family help Bryer and Emma face saying goodbye to Jack with more peace than fear?
5. At the end, Donna says Jack gave Bryer something no ribbon ever could. What do you think Jack left behind in Bryer, and what has a person or animal you love taught you?

WORDS TO KNOW

nicker the soft, low sound a horse makes to greet someone it trusts
paddock a small fenced area of ground where horses are kept
muzzle the soft front part of a horse's nose and mouth
supplement extra nutrition added to an animal's feed to help it grow strong
arthritis a condition that makes the joints swell and ache, so moving hurts

TRY THIS

Most people saw Jack as a horse with one eye and a hard past. Bryer saw a friend worth loving. Think of someone or something that others tend to overlook. Write down three good things about them that others might miss, then find a kind way this week to share one of those things.

www.ingramcontent.com/pod-product-compliance
Lightning Source LLC
LaVergne TN
LVHW090425210926
848480LV00002B/694

* 9 7 9 8 9 0 5 2 3 0 2 2 6 *